BABY ANIMALS

BABY LIONS

by Julie Murray

Cody Koala
An Imprint of Pop!
popbooksonline.com

Hello! My name is Cody Koala

This book is filled with videos, puzzles, games, and more! Scan the QR codes* while you read, or visit the website below to make this book pop.

popbooksonline.com/baby-lions

*Scanning QR codes requires a web-enabled smart device with a QR code reader app and a camera.

abdobooks.com

Published by Pop!, a division of ABDO, PO Box 398166, Minneapolis, Minnesota 55439. Copyright ©2024 by Abdo Consulting Group, Inc. International copyrights reserved in all countries. No part of this book may be reproduced in any form without written permission from the publisher. Cody Koala™ is a trademark and logo of Pop!.

Printed in the United States of America, North Mankato, Minnesota.
102023
012024

Cover Photo: Shutterstock Images
Interior Photos: Shutterstock Images; Getty Images
Editors: Elizabeth Andrews and Grace Hansen
Series Designer: Candice Keimig

Library of Congress Control Number: 2023938792

Publisher's Cataloging-in-Publication Data
Names: Murray, Julie, author.
Title: Baby lions / by Julie Murray
Description: Minneapolis, Minnesota : Pop!, 2024 | Series: Baby animals | Includes online resources and index.
Identifiers: ISBN 9781098245245 (lib. bdg.) | ISBN 9781098245801 (ebook)
Subjects: LCSH: Animal babies--Juvenile literature. | Animals--Infancy--Juvenile literature. | Lion--Juvenile literature. | Phantom cats--Juvenile literature. | Lion--Behavior--Juvenile literature.
Classification: DDC 591.39--dc23

Table of Contents

Chapter 1
Lion Cubs 4

Chapter 2
Milk to Meat. 12

Chapter 3
Growing Up 14

Chapter 4
Living in a Pride 18

Making Connections 22
Glossary .23
Index. 24
Online Resources 24

Chapter 1

Lion Cubs

Baby lions are called cubs. Pregnant females leave the **pride** to give birth. A female lion has one to six cubs at a time. Cubs are born blind and helpless.

Watch a video here!

Lion cubs weigh about three pounds (1.36kg) at birth. Their eyes are blue-gray in

color. They have spots on their fur. Over time their eyes turn brown and their spots fade.

The mother protects her cubs from **predators**. She hides them in the tall grass. She moves them every few days. She carries them in her mouth.

> Most lions live on the grasslands and open **savannas** of Africa. A small number of Asiatic lions live in the Gir Forest National Park in India.

Cubs begin to walk when they are about one month old. Mothers return to the

pride with their cubs when the cubs are about eight weeks old.

Chapter 2

Milk to Meat

Cubs drink their mother's milk. Over time, they also drink milk from other females in the **pride**. They start eating meat when they are two to three months old.

Learn more here!

Chapter 3

Growing Up

Lion cubs love to run and wrestle. As they grow, females teach them to hunt and survive in the wild. Cubs begin to roar when they are one year old.

A lion's roar can be heard five miles (8km) away.

Explore links here!

Male cubs leave the **pride** when they are two to three years old. Female cubs stay

with the pride for life. Lions are fully grown when they are about five years old.

Chapter 4
Living in a Pride

A **pride** can have up to 40 members. It is made up of a few males, many females, and their young. Lions are the only cats that live in large, **social** groups.

Where Lions Live

Complete an activity here!

Females in a pride hunt together. They hunt zebras, impala, buffalo, and

Lions can live about 15 years in the wild.

other grassland animals. Males protect their pride and **territory**.

Making Connections

Text-to-Self

What is one new fact you learned about lion cubs? Why did you find that fact interesting?

Text-to-Text

Have you read a book or seen a movie that has a lion cub as a character? How was that lion cub similar to or different from the lion cubs talked about in this book?

Text-to-World

Lions live in groups called prides. What other animals live in groups?

Glossary

predator – an animal that hunts other animals for food.

pride – a family unit of lions. A pride can include up to three males, 12 females, and their young.

savanna – a tropical grassland with scattered trees.

social – living in groups or communities instead of alone.

territory – a particular area of land that an animal or group of animals claims.

Index

behaviors, 4, 9–12, 14, 16–18, 20–21

birth, 4, 6

eyes, 6–7

food, 12, 20–21

fur, 7

hunting, 14, 20

markings, 7

prides, 4, 11–12, 16–18, 20–21

sounds, 14

weight, 6

Online Resources

popbooksonline.com

Thanks for reading this Cody Koala book!

This book is filled with videos, puzzles, games, and more! Scan the QR codes* while you read, or visit the website below to make this book pop.

popbooksonline.com/baby-lions

*Scanning QR codes requires a web-enabled smart device with a QR code reader app and a camera.